THE GIST
OF CALCULUS

by

Borislav Dzodzo

Contents

THE GIST OF CALCULUS

BY

Борислав М. Цоцо

INTRODUCTION

The front page features the photogenic K9 that goes by the name of Orion, whose depiction was captured by his owner and talented photographer and architect Melissa Thereliz . I would like to thank Dr Relja Vulanovic, Professor, Kent State University at Stark, USA who ensured that there are far fewer mistakes in this book. If you are actually bothering to read this book it is likely that the traditional approach to calculus education confused you as much as it confused me. Calculus can be taught in many different ways and the geometric approach made the most sense to me but I never encountered it in my course of study. I was taught the detailed minutia of calculus in my classes but the geometry that underlies the core of Calculus was never be explained to me. I was simply given formulas that seemed to have descended from the heavens and was taught to eventually accept them on good faith as an unquestionable truth and then told to perform operations with them in hopes that somchow I will eventually understand calculus by memorizing these formulas. This encounter with a faith based approach to calculus left me with a lingering sense that the math department has betrayed the principles of reason upon which the mathematics was built. The stunning simplicity and beauty of the basic derivatives was cruelly hidden from me at the time. This book will try to give you a visual representation of what some of the basic derivatives are all about. Your regular classes and the books used to teach them will provide you with all of the practical applications, quiz material, homework questions and other far more boring bits that belong in a proper Calculus book. This book strictly concerns itself with the core ideas that form the geometric backbone of calculus. Since calculus is an old topic studied by millions of brilliant minds I doubt that there are any new or recent proofs within these pages so I feel no particular need to take credit for the ideas or bother you with the details of people who are believed to have been the first to think of them.

DERIVATIVE

What is a derivative anyway? Derivatives are functions that describe how quickly other functions change. If a function around some value of x moves up in value quickly as the values of x grows then it could be said that the derivative of this function is large and positive around those values of x. If a function stays about the same as x grows around some neighbourhood of x values then it could be said that the derivative function stays close to zero around those values of x. If for some neighbourhood of x values the function decreases in value as x grows then the derivative of such a function is negative in that neighbourhood of x values. You might be concerned that in a neighbourhood of x the function doesn't just go up or down but it has hills and valleys of all sorts. To keep things simple we will only consider functions such that if we choose a narrow enough neighborhood of x then the values of the function only go up or down. The neighborhood that we will consider will be so very tiny that within it the function itself will loose almost all of its curviness and become almost a completely straight line.

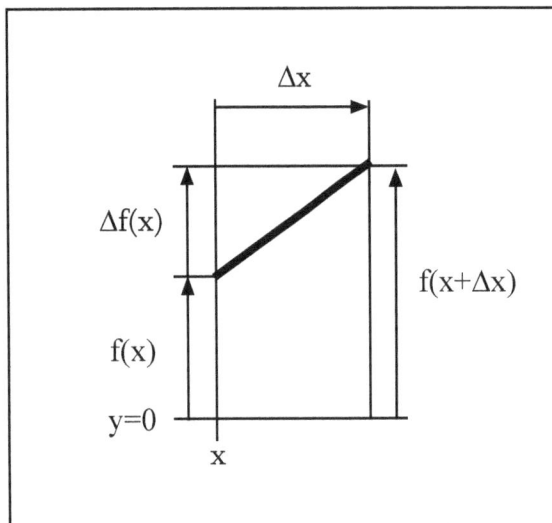

Fig. 1 f(x) in x neighbourhood

If we take a look at Fig. 1 we can see the neighborhood of the function f(x) at some value x. The size of the neighborhood will be Δx in width. You may notice that as long as we keep our distances very small it is true that if you increase the size of Δx the size of Δf(x) also increases proportionally. For this reason we can't simply talk about the size of Δf(x) but we must consider the proportion of the size of Δf(x) to the size of Δx. This is another way of saying that we are interested in the ratio Δf(x)/Δx which is also called the rate of change.

We could consider all kinds of other relationships between these two values such as $\Delta x/\Delta f(x)$ or $\arctan(\Delta f(x)/\Delta x)$ or some other relationship but the most useful measure of the slope in the later applications of calculus happens to be $\Delta f(x)/\Delta x$. If we make the two numbers infinitesimally small then all that we are left with is the exact rate of change at that particular position which is going to be called the derivative and the function that provides the value of the derivative at the position x is written as f '(x). How do we know that the ratio doesn't just disappear and become zero or infinity as the Δx and $\Delta f(x)$ are shrunk? This has to do with something called limits that a more formal calculus book covers in greater detail but for us it is sufficient to imagine that the two sides of a triangle describing the rate of change have the same proportions to each other regardless of how small the triangle itself becomes.

If: f(x) has a derivative

Then: f '(x) = $\Delta f(x)/\Delta x$
Where $\Delta f(x)$ and Δx are infinitesimally small at each particular value of x

ADDITION

If we have two functions whose sum composes the third function, can we determine the derivative of the third function if we know the derivatives of the functions that compose it?

Let us say we have: $f(x) = g(x) + r(x)$

We are looking for f'(x) and we already know what the graphs of g'(x) and r'(x) look like. In this case the derivative of g(x) function is simply added to the derivative of r(x) function to get the derivative of f(x) function.

$$f'(x) = g'(x) + r'(x)$$

The reason for this is simple. To make it more obvious let us consider what does the derivative of the following function look like.

$$f(x) = g(x) + g(x)$$

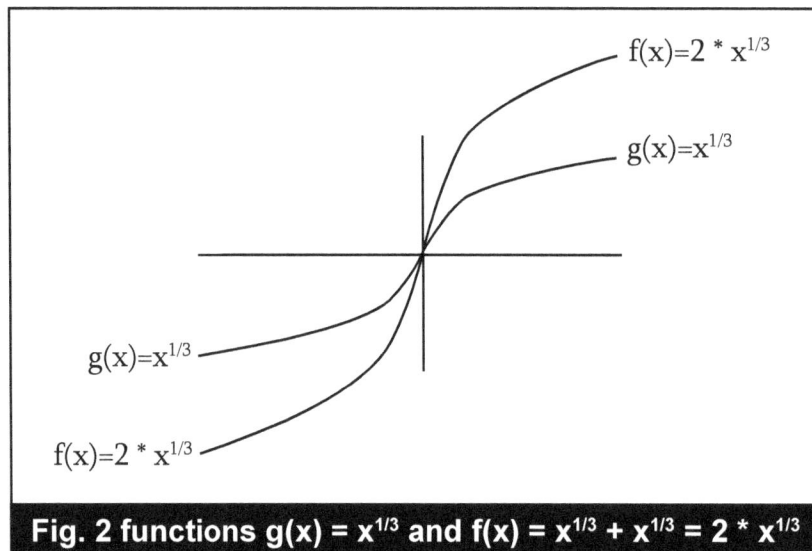

$f(x)=2 * x^{1/3}$

$g(x)=x^{1/3}$

$g(x)=x^{1/3}$

$f(x)=2 * x^{1/3}$

Fig. 2 functions g(x) = x$^{1/3}$ and f(x) = x$^{1/3}$ + x$^{1/3}$ = 2 * x$^{1/3}$

In the Fig. 2 the function $f(x)=x^{1/3} + x^{1/3}$ is equal to the sum of the two $g(x)=x^{1/3}$ functions. The derivative of $f(x)$ is not the same as that of $g(x)$. The derivative of $f(x)$ has double the rate of change of $g(x)$ precisely because the function $f(x)$ looks to be a vertically stretched out version of $g(x)$ so that each slope of $f(x)$ goes up or down twice as much as the slope of $g(x)$. Because of this we can say that

$$f'(x) = g'(x) + g'(x) = 2\,g'(x)$$

We can generalize this principle and also say that for a function $f(x) = C * g(x)$ the derivative will be equal to the derivative of $g(x)$ multiplied by the constant C. The derivative of $f(x) = C * g(x)$ is equal to $f'(x) = C * g'(x)$.

If we return to the general case of adding two very different functions the resulting function will not necessarily look like a vertically stretched version of ether one of the original functions.

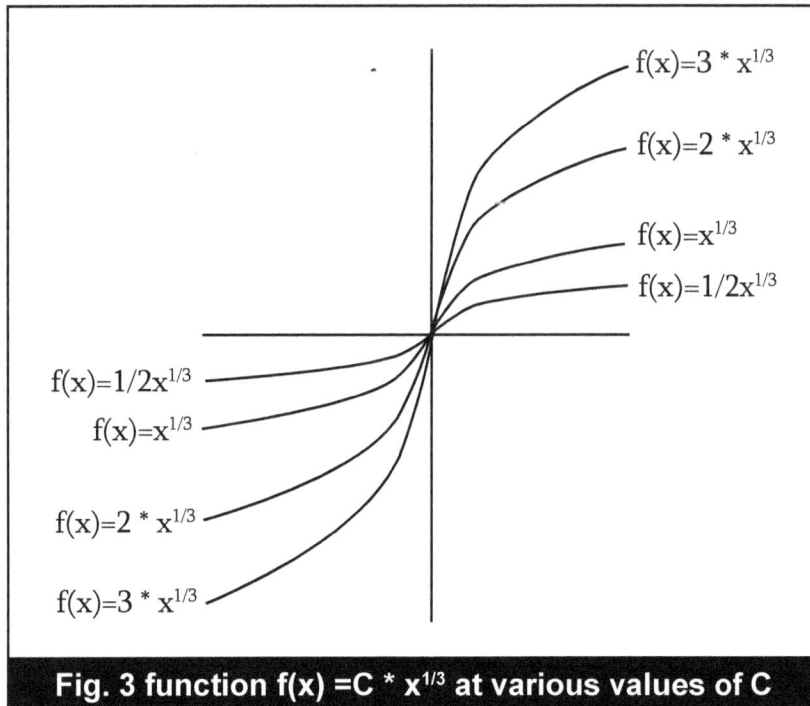

Fig. 3 function f(x) =C * x$^{1/3}$ at various values of C

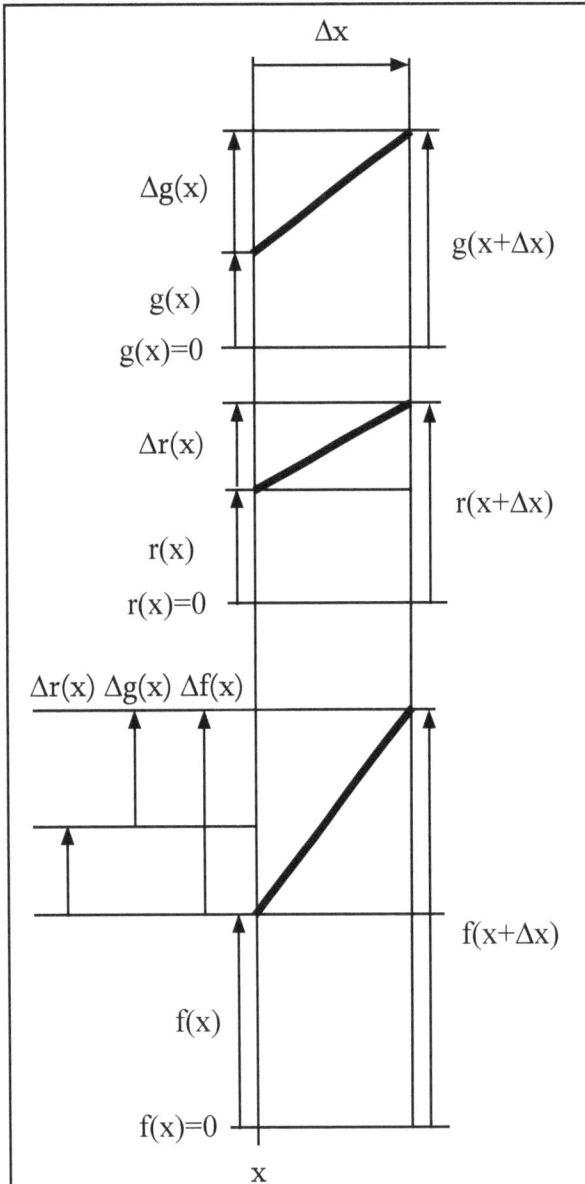

Δx

$\Delta g(x)$

$g(x+\Delta x)$

$g(x)$

$g(x)=0$

$\Delta r(x)$

$r(x+\Delta x)$

$r(x)$

$r(x)=0$

$\Delta r(x)$ $\Delta g(x)$ $\Delta f(x)$

$f(x+\Delta x)$

$f(x)$

$f(x)=0$

x

Fig. 4 g(x), r(x) and f(x) at some value of x

This is an opportune time to pull out our mathematical microscope and zoom into the neighbourhood of the two functions that will be at some value of x (see Fig. 4). Please notice that when you zoom in far enough the functions lose their splendid rolling curves and become simple almost linear slopes. If this isn't the case with our particular function then the situation becomes too complicated for the scope of this book but it is usually explained in the regular Calculus class.

From Fig. 4 we can see that

$$g(x + \Delta x) = g(x) + \Delta g(x)$$

It can also be seen that

$$r(x + \Delta x) = r(x) + \Delta r(x)$$

We know that

$$f(x) = g(x) + r(x)$$

$$f(x + \Delta x) = g(x + \Delta x) + r(x + \Delta x)$$

$$f(x + \Delta x) = f(x) + \Delta f(x)$$

Since we know that $f(x + \Delta x) = f(x) + \Delta f(x)$

And we also know $f(x + \Delta x) = g(x) + \Delta g(x) + r(x) + \Delta r(x)$

Which is rewritten as $f(x + \Delta x) = g(x) + r(x) + \Delta g(x) + \Delta r(x)$

Substituting variables $f(x + \Delta x) = f(x) + \Delta g(x) + \Delta r(x)$

Subtracting by $f(x)$ $f(x + \Delta x) - f(x) = \Delta g(x) + \Delta r(x)$

Since $f(x + \Delta x) - f(x) = \Delta f(x)$ $\Delta f(x) = \Delta g(x) + \Delta r(x)$

That is to say that the small vertical change in the sum function $f(x)$ is equal to the sum of the two small vertical changes in the functions $g(x)$ and $r(x)$.

Since the derivative compares the change in the function to the change in x values we can divide both sides of the previous equation with Δx in order to get the rate of change and we have:

$$\Delta f(x)/\Delta x = \Delta g(x)/\Delta x + \Delta r(x)/\Delta x$$

If we make the change of x really small then we get the relationship of the derivatives which is:

$$\text{If: } \quad f(x) = g(x) + r(x)$$

$$\text{Then: } \quad f'(x) = g'(x) + r'(x)$$

MULTIPLICATION

What does a derivative of a function look like if it is composed of two multiplied functions whose derivatives are known. That is to say if we have $f(x) = g(x) * r(x)$ and we know $g'(x)$ and $r'(x)$ then what could the $f'(x)$ possibly be?

The easiest way to understand this is to have a mental model of what multiplication really represents. You may remember a fact that an area of a rectangle is equal to the product of its two side lengths. This is a very useful analogy to work with if we are to understand how the formula for a derivative of two multiplied functions comes about. Please note in Fig. 5 that the rectangle representing $f(x)$ grows a really small amount because both $g(x)$ and $r(x)$ get bigger by Δg and Δr respectively.

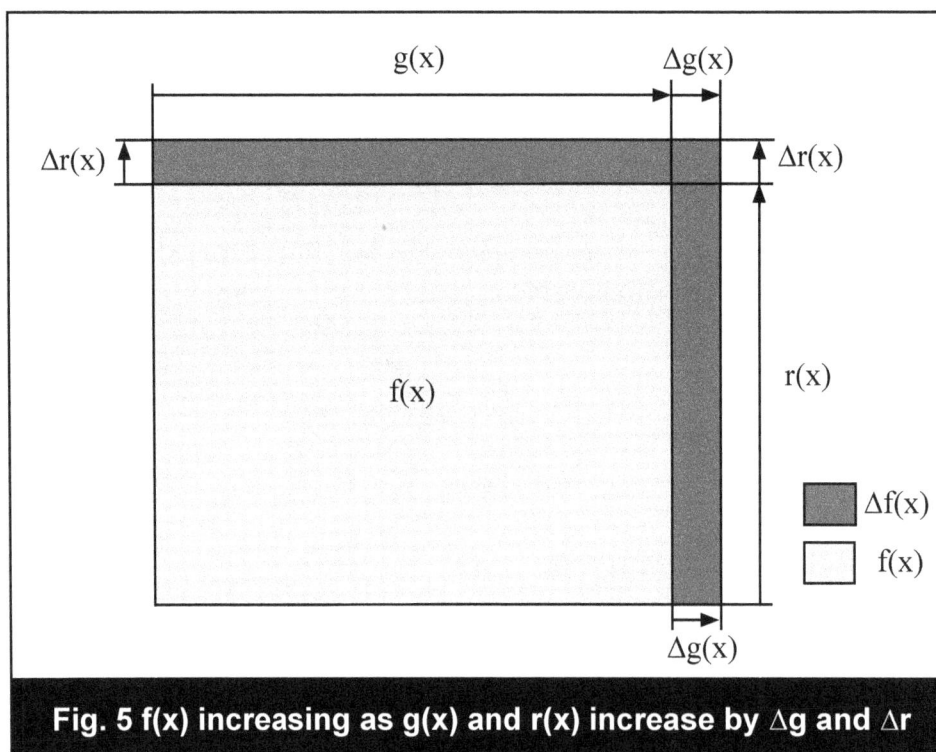

Fig. 5 f(x) increasing as g(x) and r(x) increase by Δg and Δr

Notice in Fig. 5 that if only the function g(x) grows by a small amount Δg(x) then we add a small rectangle to the existing area representing f(x). The formula for the rectangle growth in that case is as follows:

$$\Delta f(x) = \Delta g(x) * r(x)`$$

If the function r(x) is the only one that changes then the formula is:

$$\Delta f(x) = g(x) * \Delta r(x)$$

If both of the functions change then the formula is

$$\Delta f(x) = \Delta g(x) * r(x) + g(x) * \Delta r(x) + \Delta g(x) * \Delta r(x)$$

If we want to find out the rate of change of f(x) then we can't simply consider the change in the function f(x) without taking into account the related change in x. For this reason we divide both sides by the Δx and we get.

$$\Delta f(x)/\Delta x = \Delta g(x)/\Delta x * r(x) + g(x) * \Delta r(x)/\Delta x + \Delta g(x) * \Delta r(x) / \Delta x$$

When the Δx becomes very small then we get

$$f'(x) = g'(x) * r(x) + g(x) * r'(x) + a \text{ problem leftover :)}$$

The problem leftover can be written as

$$g'(x) * \Delta r(x)$$
or
$$r'(x) * \Delta g(x)$$
or
$$(\Delta g(x) / \Delta x) * (\Delta r(x) / \Delta x) * \Delta x \approx g'(x) * r'(x) * \Delta x$$

If g(x) and r(x) are functions which have derivatives then their derivatives are some finite numbers between positive and negative infinity. This means that g'(x) or r'(x) is multiplied with a number that tends toward zero as Δx is made smaller and so the whole problem leftover ultimately becomes zero.

To make the mathematics of the problem leftover slightly clearer let us observe the difference in what takes place when x tends towards zero in functions $y = \Delta x / \Delta x$ and $y = \Delta x * \Delta x / \Delta x$. The equation $y = \Delta x / \Delta x$ turns into 1 as Δx approaches zero because the numerator and denominator balance each other out. In contrast the function $y = \Delta x * \Delta x / \Delta x$ becomes zero as Δx comes closer to zero. Since this function can be represented as $y = 1 * \Delta x$ and for this reason it is clear why it becomes zero as x tends towards zero. This will be explained in a more formal way and in greater detail in a proper calculus book but just keep this explanation in mind as a stand in for a more formal one.

Since the problem leftover becomes too small too quickly to stay relevant as Δx gets smaller, the end result of our simplification and elimination of leftover terms gives us the final result which states that:

$$\text{If: } f(x) = g(x) * r(x)$$

$$\text{Then: } f'(x) = g'(x) * r(x) + g(x) * r'(x)$$

DIVISION

In case of division we will try to find a visual analogy to describe what division really represents. While a rectangle was a very useful analogy for multiplication we will use a pair of similar triangles to represent division.

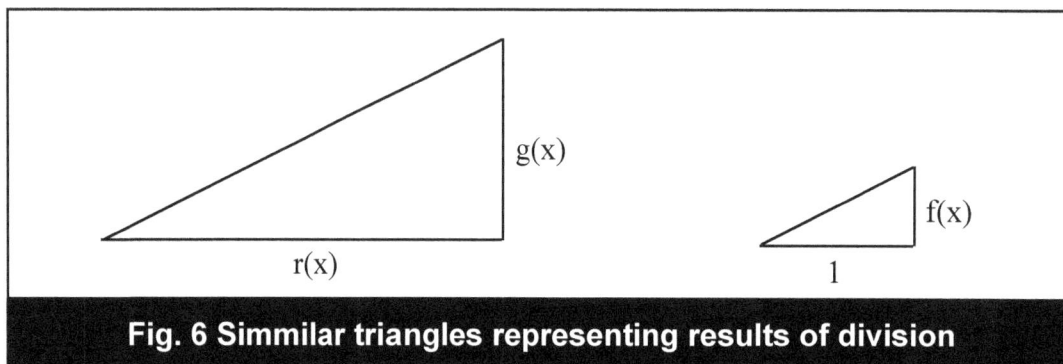

Fig. 6 Simmilar triangles representing results of division

The relationship f(x)/1 = g(x)/r(x) can be represented geometrically with two similar triangles. One side of the resulting triangle has to be equal to one unit. The triangles that we will use will have a right angle so that the effect of growth of the numerator and denominator can be easily understood.

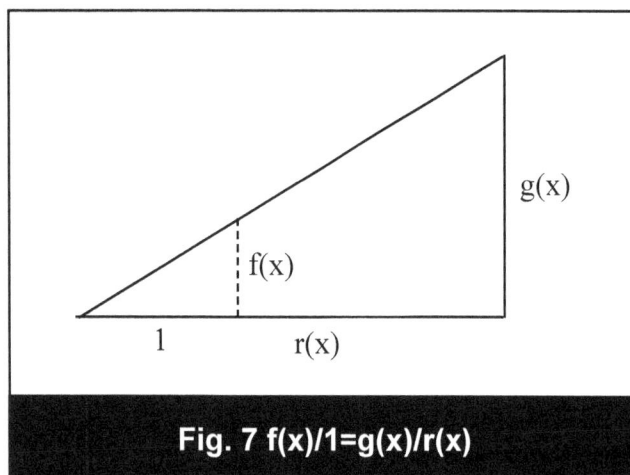

Fig. 7 f(x)/1=g(x)/r(x)

Because Fig. 6 depicts similar triangles we can say

$$\frac{f(x)}{1} = \frac{g(x)}{r(x)}$$

or simply
$$f(x) = g(x)/r(x)$$

Fig. 7 makes this relationship more obvious by placing the smaller result triangle inside the larger triangle representing the division.

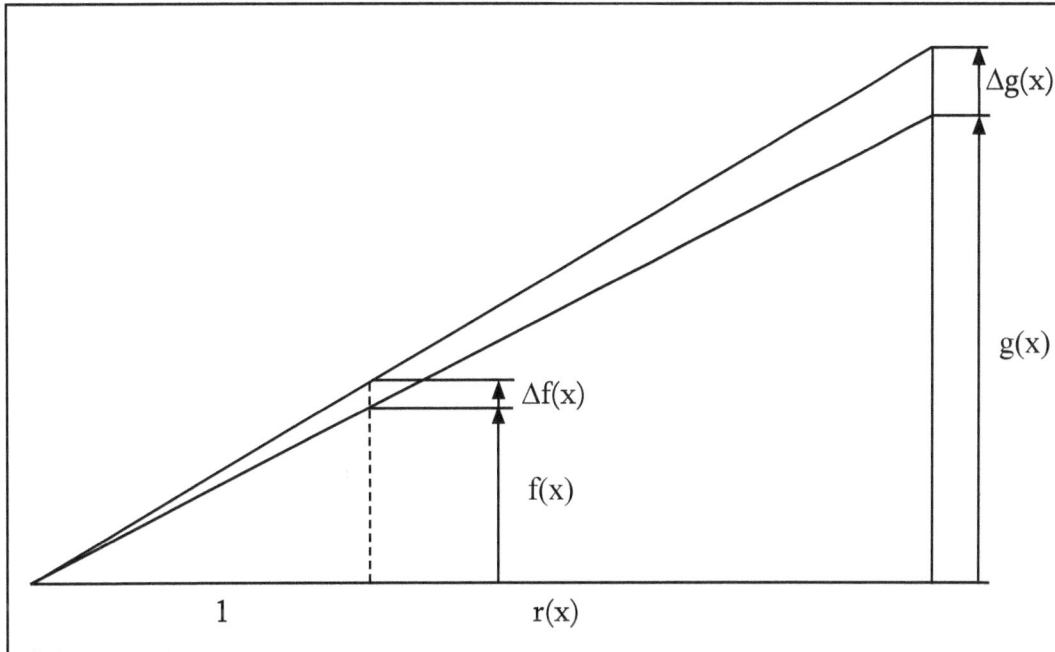

Fig. 8 The effect of Δg(x) on Δf(x)

Let us observe in Fig. 8 what kind of an effect does a change in g(x) have on the change in f(x). If we carefully analyze the diagram it will be noticed that if only g(x) changes then f(x) will change proportionally. The relationship of ratios for slightly larger triangles is as follows.

$$\frac{f(x) + \Delta f(x)}{1} = \frac{g(x) + \Delta g(x)}{r(x)}$$

Subtracting out the ratios of smaller triangles on each side.

$$\frac{f(x) + \Delta f(x)}{1} - \frac{f(x)}{1} = \frac{g(x) + \Delta g(x)}{r(x)} - \frac{g(x)}{r(x)}$$

We get a simplified relationship which should still be obvious in Fig. 8

$$\frac{\Delta f(x)}{1} = \frac{\Delta g(x)}{r(x)}$$

$$\Delta f(x) = \Delta g(x)/r(x)$$

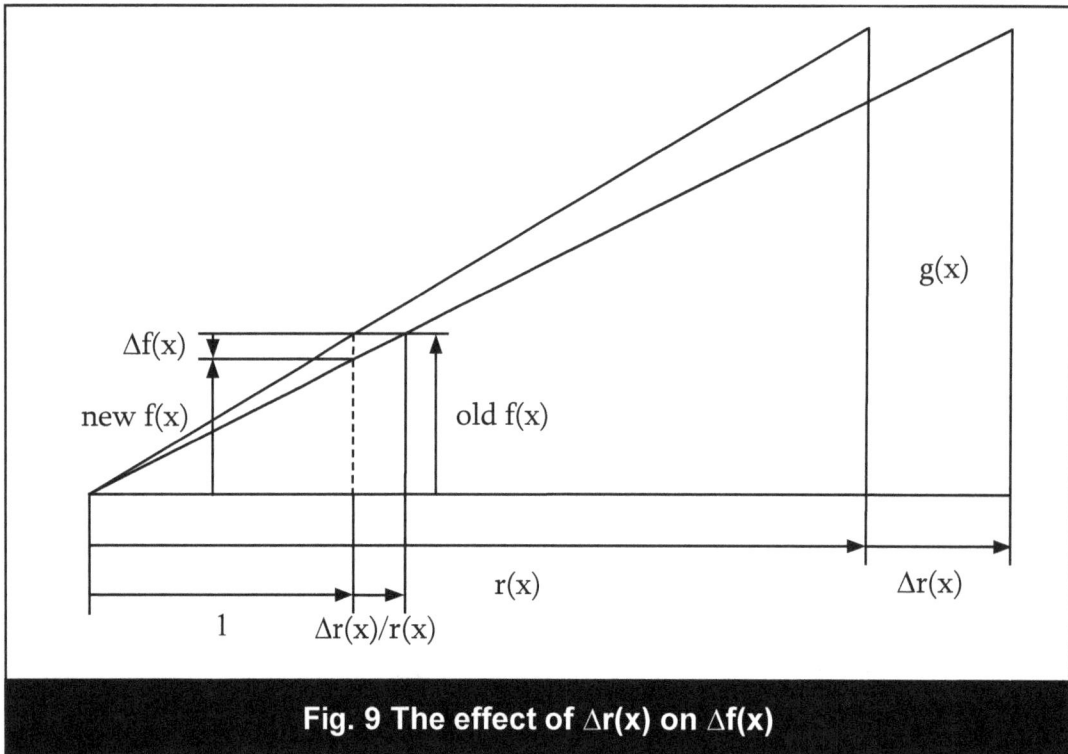

Fig. 9 The effect of Δr(x) on Δf(x)

Fig. 9 demonstrates how a change in r(x) has a much different effect on the change in f(x). As r(x) becomes larger the value of f(x) becomes smaller. This makes sense because what is really taking place is that we are dividing g(x) by a greater number. What stands to be determined is the exact amount by which f(x) becomes smaller as r(x) increases.

Fig. 10 shows the stretching of the original unit triangle which represents the result of the division. Using the proportions of similar triangles we can say

$$\frac{\Delta f(x)}{\Delta r(x)/r(x)} = -1 * \frac{old\ f(x)}{1 + \Delta r(x)/r(x)}$$

Which can also be written as

$$\frac{\Delta f(x)}{old\ f(x)} = -1 * \frac{\Delta r(x)/r(x)}{1 + \Delta r(x)/r(x)}$$

$\Delta f(x)$

new f(x)

old f(x)

1 $\Delta r(x)/r(x)$

what was 1 unit long before $\Delta r(x)$ was added to r(x)

β $\Delta f(x)$

α

β $\dfrac{\Delta r(x)}{r(x)}$

new f(x)

old f(x)

α

1 $\Delta r(x)/r(x)$

what was 1 unit long before $\Delta r(x)$ was added to r(x)

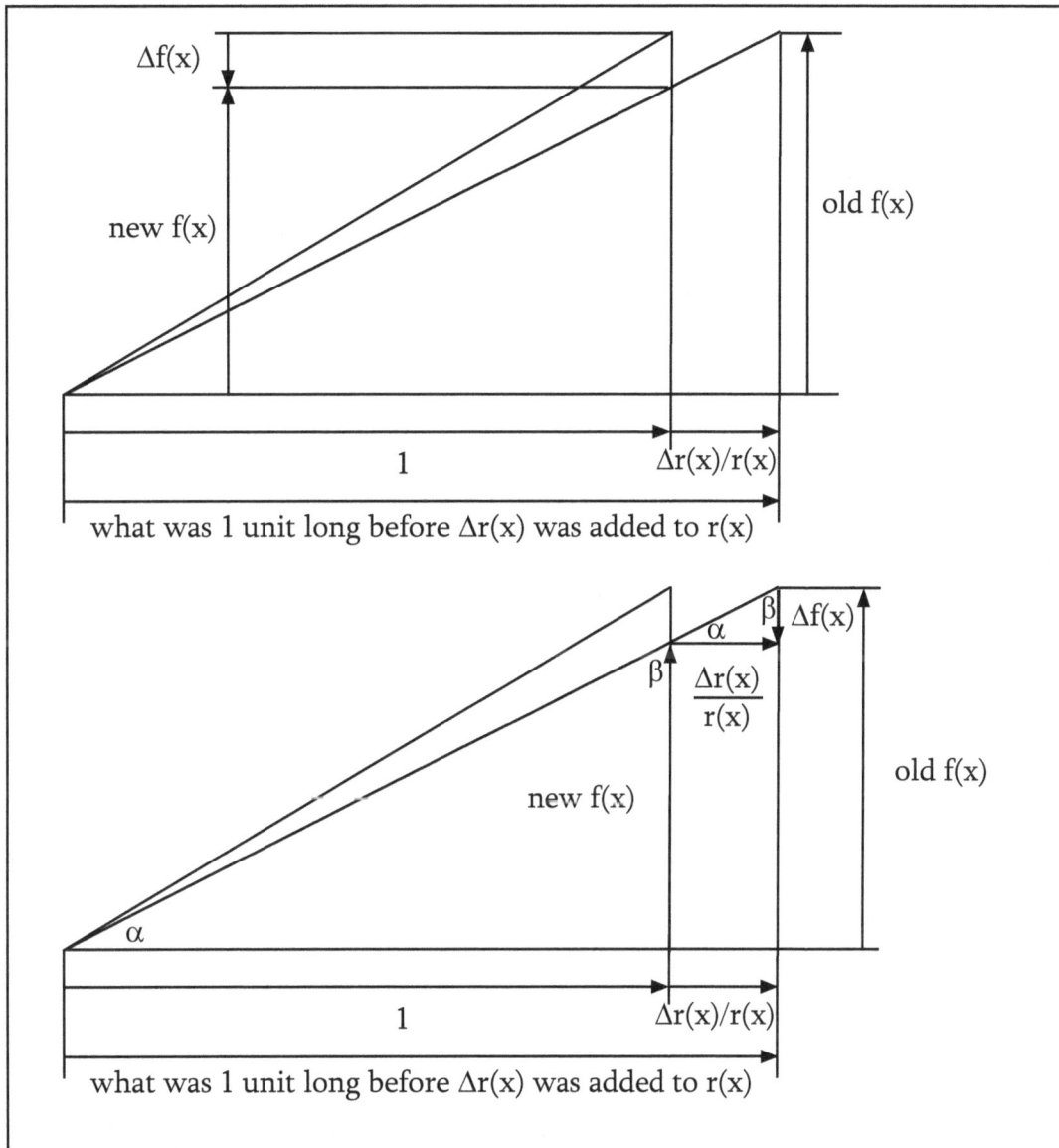

Fig. 10 Detailed view of the effect of $\Delta r(x)$ on $\Delta f(x)$

Since we know that the old f(x) is actually function g(x)/r(x) we can say

$$\Delta f(x) = \frac{g(x)}{r(x)} * \frac{-\Delta r(x)/r(x)}{1 + \Delta r(x)/r(x)}$$

$$\Delta f(x) = \frac{g(x)}{r(x)} * \frac{-\Delta r(x)}{r(x) + \Delta r(x)}$$

$$\Delta f(x) = \frac{g(x)}{r(x)} * \frac{-\Delta r(x)}{r(x) + \Delta r(x)}$$

If we combine the effects of changing g(x) and r(x) we get

$$\Delta f(x) = \frac{\Delta g(x)}{r(x)} - \frac{\Delta r(x) * g(x)}{r(x) * (r(x) + \Delta r(x))}$$

$$\Delta f(x) = \frac{\Delta g(x) * (r(x) + \Delta r(x))}{r(x) * (r(x) + \Delta r(x))} - \frac{\Delta r(x) * g(x)}{r(x) * (r(x) + \Delta r(x))}$$

which can be simplified further

$$\Delta f(x) = \frac{\Delta g(x) * r(x) + \Delta g(x) * \Delta r(x) - \Delta r(x) * g(x)}{r(x) * r(x) + r(x) * \Delta r(x)}$$

What really interests us is the rate of change in f(x) proportional to the change in x. The multiplication example clearly demonstrated that in order to do this we need to divide both sides by the change in x in order to get

$$\Delta f(x)/\Delta x = \frac{(\Delta g(x)/\Delta x) * r(x) + \Delta g(x) * \Delta r(x) / \Delta x - (\Delta r(x)/\Delta x) * g(x)}{r(x) * r(x) + \Delta r(x) * r(x)}$$

In order to find out the exact rate of change of f(x) in a particular position we need to make the change in x really really small. If we make the change in x infinitesimally small then this complicated relationship simplifies and becomes

$$f'(x) = \frac{g'(x) * r(x) + \text{problem leftover} - r'(x)\, g(x)}{r(x) * r(x)}$$

The problem leftover is the same as the one in the multiplication example and it turns into zero as the change in x becomes really small for the same reason that it does in the multiplication example.

Part of the denominator $\Delta r(x) * r(x) = (\Delta r(x) / \Delta x) * r(x) * \Delta x = r'(x) * r(x) * \Delta x$ becomes zero if Δx is infinitesimally small and for this reason it disappears from the equation as well.

Finally it can be concluded that:

If: $f(x) = g(x)/r(x)$

Then: $f'(x) = \dfrac{g'(x) * r(x) - g(x) * r'(x)}{r(x)^2}$

FUNCTION IN A FUNCTION

If we know the derivative of g(x) and r(x) would it be possible to know the derivative of f(x) if f(x) = g(r(x)) ? In order to answer this question I will implore you to use the power of your imagination. Let us imagine that we are looking at a ball rolling across the street in front of us. We can see that the motion of the ball is constant and so we might say that the ball position is equal to some constant C multiplied by time. More formally it could be written as p = C * x where x represents time and this will be our function g(x). The rate of change of the ball movement, or rather its derivative in this case is simply equal to C. At this point we shall engage our imagination 100% and pull out of our pockets a device which slows down time so that everything looks like it is moving in slow motion. We shall turn the dial to slow down world time by a factor of two. More formally we can write this down as World Time = 1/2 * Our Time and this can be our function r(x). The derivative of such a function is 1/2. We know that this would slow down the ball movement by twice as much. Formally stated the relationship describing the position of the ball in our time happens to be p = C * 1/2 * Our Time or g(r (x)). The derivative of this function is C * 1/2. This example seems to imply in the situations where the derivative of g(r(x)) would be equal to g'(x) * r'(x). This seems too simple and at this point it is very questionable if this relationship generalizes to all functions.

The first concern is that these functions are simple and linear and not some complicated functions with many hills and valleys.   To address this we must understand that when we are looking for a derivative at a particular point then all of the complexity of the function and the shape of the curve in other locations doesn't really matter. If we zoom into the local neighbourhood the function should have an appearance of a nearly straight sloped line. The derivative is the measure of how sloped the line is at that point. If we are watching the ball drop from a building instead of roll across the street we will notice that the ball picks up speed as it falls and its derivative (the rate of change of its position) is different at different times. If we can slow down time by a factor of two however and then zoom in to inspect how this effects each neighbourhood along the modified curve representing the change of the ball position over time we will notice that each slope is now twice as shallow and for this reason the derivative of the final function is twice as small. Even if the curve describing the motion is far more complex then the one describing a falling ball we could still say that the derivatives describing the motion of such an object would be half as large if we could slow down time by a factor of two. It must be noted that a problem with our previous assumption becomes apparent in this example. The previous example led us to conclude that f '(x) = g'(x) * r'(x) however this was concluded in a situation where both g'(x) and r'(x) remain constant for all values of x. In this example g'(x) changes with time and since we are looking at the rate of change of g(x) at our point in time which has gone twice as far ahead of the world time is not the point in time that the ball is actually experiencing. We should disregard the derivative g'(x) which tells us the rate of change of the ball position at our time and instead consider g'(r(x)) derivative that describes how the ball position changes at a point in world time that corresponds to our point in time. This means that the derivative of f '(x) is actually now f '(x) = g'(r(x)) * r'(x). Although we allowed the function describing the position of the object to become complicated we still kept the function describing the change between our time and world time to be very simple and its derivative is the same at all times.

If we complicate the mechanism by which we alter the motion of time and time doesn't simply slow down but maybe slows down at an increasing rate or slows down and then speeds up or maybe even goes backwards then we have a more general description of the function r(x). It is important to note that regardless of what our time altering function looks like we will consider what effect it has by zooming in to examine its time altering properties at a particular moment and we will find out that if we zoom in far enough the change in how the time is slowing down, speeding up or going backwards is linear. That is to say that if we are looking from the perspective of our time we can see that the world time is moving at a different and steady pace and that in the very near future point of our time the world time will also change by a small amount. If we zoom in far enough we should even be able to say that these relationships are nearly linear. This means that when we zoom in, then for a small change in our time, say 2 microseconds, the world time at that point will change by 5 microsecond. If the small change in our time is twice as big say 4 microseconds then the world time should change by 10 microseconds. At a different point in our time this relationship could easily be different, say 3 microseconds of our time is 1 microsecond of world time or even -1 microsecond of world time if the world has started to move back in time. This relationship is in a sense described by the derivative function, so what we are saying is that at every point in our time the derivative of the time changing function will describe how fast the world time is going. We already know from the previous paragraph what effect does the time slowdown or speedup have on the derivative of the function describing the object position. If the speed of world time at a particular moment is twice that of our speed then the object will move twice as fast, and if the speed of world time is only half as fast as ours then the object position will move twice as slow. If the speed of time is negative, which is to say that the time is going backwards, then the object will move backwards and the change of position will be in the opposite direction of what it "should" be. In all of these examples it is important to recognize that the speed of time, represented by the derivative of r(x) is multiplied by the rate of change of the position of the object represented by the derivative of g(x) at the relevant point in world time given by r(x). Since the derivative of our time change function tells us how quickly time is changing in the world at a particular point in our time, we can simply multiply that value with the change in position of the object at the corresponding point in world time in order to find out the change in position at a particular point in our time. This is another way of saying that in general if we have f(x) = g(r(x)) then at any of point of x we can multiply g'(r(x)) by r'(x) in order to find out f '(x).

The previous verbal description of the derivative of g(r(x)) may have been sufficient to convince you that f '(x) = g'(r(x)) * r'(x) however there is a visual way of conveying the same explanation. Fig. 11 has a description of a really small segment of what may be a complicated function r(x) around the value x. Notice that the small segment is almost linear and also note that the value of

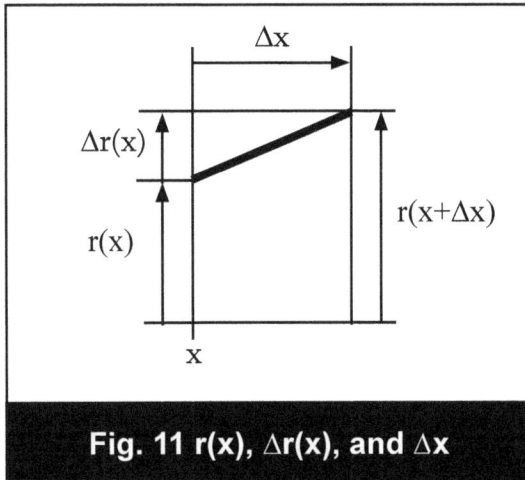

Fig. 11 r(x), Δr(x), and Δx

r(x) changes less then the value of x. Fig. 12 shows the values of g(x) in a very small neighborhood around r(x).

When we zoom into these functions we find that on the small scale the functions become lines with different slopes. We can see that the bigger the change in x the bigger the change in r(x) and the bigger the change in r(x) the bigger the change in g(r(x)).

It is important to realize that in the Fig. 12 the "x" axis is that of r(x) and not x itself. This means that positions along the r(x) axis represent regularly increasing values of r(x) just as they do on the x axes on graphs depicting functions of x. The two functions are only different if r(x) never reaches some range of values. This means that within the range where r(x) has a value the shape of the curve of g(r(x)) on the graph with the axis of r(x) would look identical to the graph of g(x) on the graph with the axis of x. This also means that the derivatives of such curves would be identical with the important difference that g'(r(x)) is a derivative in respect to the change in r(x) while g'(x) is a derivative in respect to the change in x.

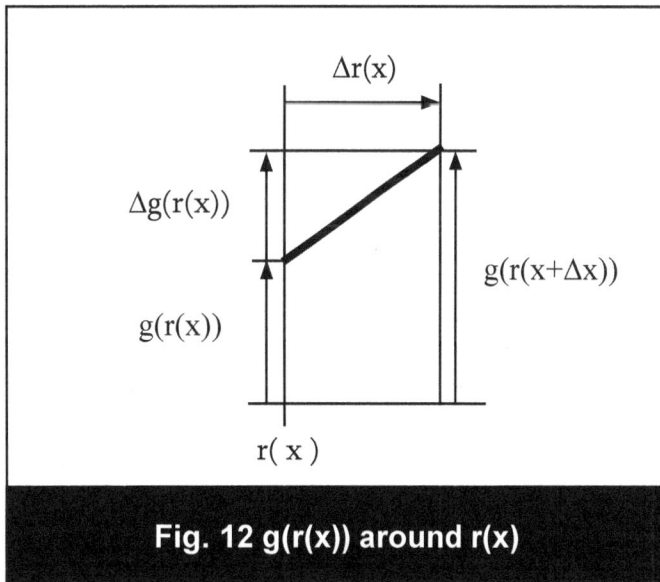

Fig. 12 g(r(x)) around r(x)

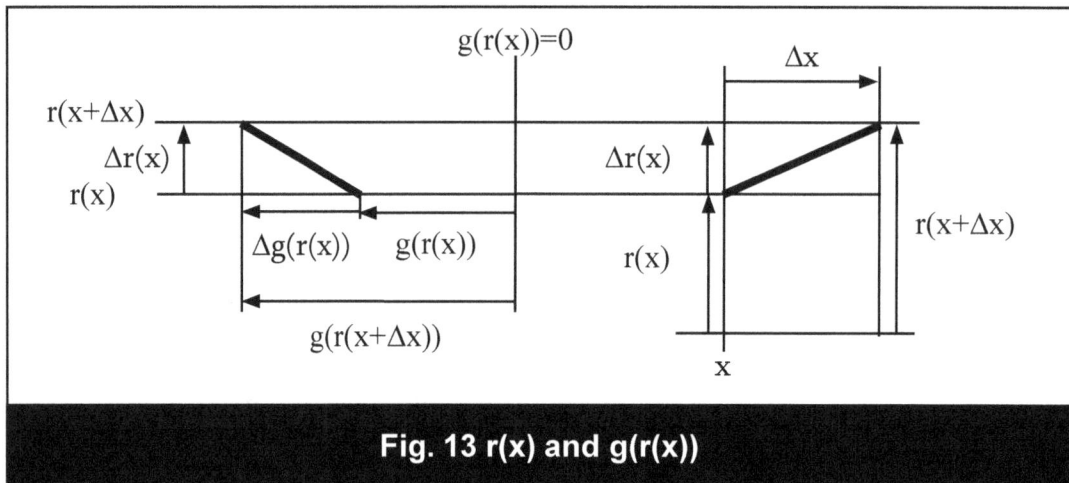

Fig. 13 r(x) and g(r(x))

The Fig. 13 depicts the rate of change of g(r(x)) around the value of r(x) which itself is determined by the value of x. In the Fig. 13 it is important to note that the change in x is now related to the change in g(r(x)) through the change in r(x). Given these proportions, Δr(x) is related to Δx by the proportion Δr(x)/Δx, and the Δg(r(x)) is related to Δr(x) by the proportion Δg(r(x))/Δr(x). It could then be concluded that the Δg(r(x)) is related to Δx by the proportion [Δg(r(x))/Δr(x)] * [Δr(x)/Δx].

Fig. 14 shows an example of how the relationship of proportions between Δg(r(x)) and Δr(x) has to be scaled by the relationship between Δr(x) and Δx in order to find the relationship between Δg(r(x)) and Δx. The proportion of Δg(r(x)) and Δr(x) is shown by a triangle where Δr(x) = 1. A simmilar triangle is presented where Δr(x) is scaled down to reflect the size of the change in r(x) when the change in Δx = 1. For that reason the Δg(r(x)) side is also scaled to be Δg(r(x))/Δx units in size. Since these relationships form two simmilar triangles it can be said that

$$\frac{\Delta g(r(x))}{\Delta r(x)} = \frac{\Delta g(r(x))/\Delta x}{\Delta r(x)/\Delta x}$$

Multiplying both sides by Δr(x)/Δx

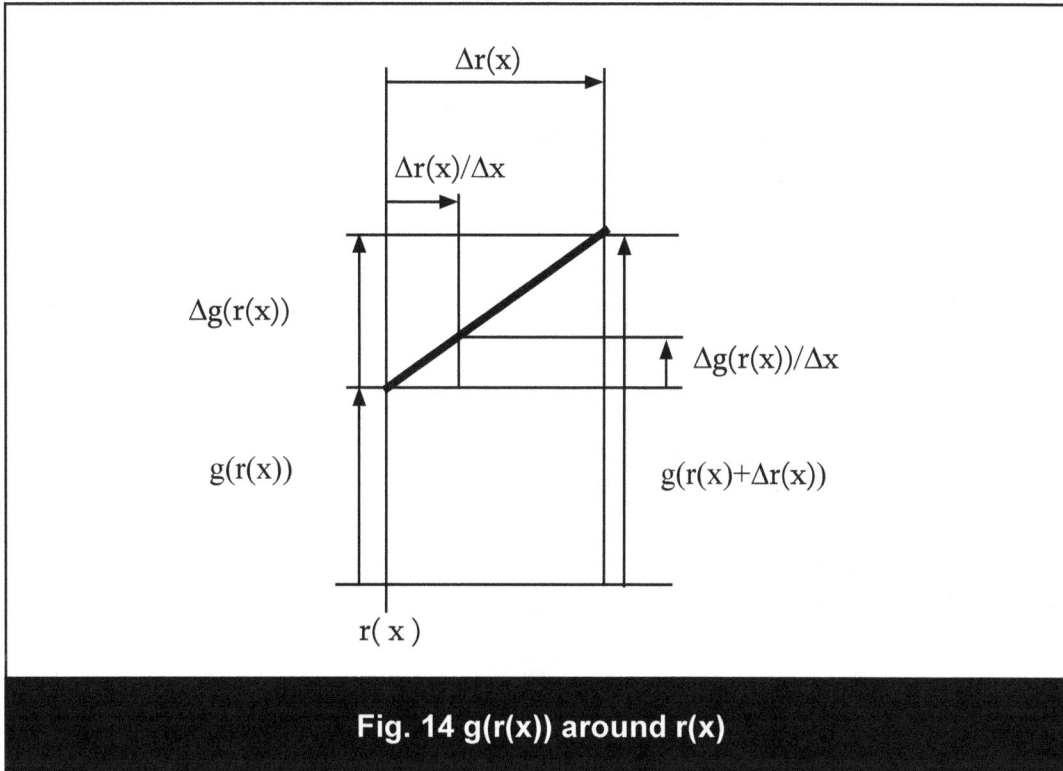

Fig. 14 g(r(x)) around r(x)

We get

$$\Delta g(r(x)) / \Delta x = [\Delta g(r(x)) / \Delta r(x)] * [\Delta r(x) / \Delta x]$$

Since $f(x) = g(r(x))$ and $\Delta f(x) = \Delta g(r(x))$ we can say

$$\Delta f(x)/\Delta x = [\Delta g(r(x))/\Delta r(x)] * [\Delta r(x)/\Delta x]$$

which turns into

$$f'(x) = g'(r(x)) * r'(x)$$

It can be concluded that:

If: $f(x)=g(r(x))$

Then: $f'(x) = g'(r(x)) * r'(x)$

FUNCTION INVERSE

If there is a function y=f(x) then there could be a function y=f⁻¹(x) such that when a number x is put into the function f⁻¹(f(x)) we get x back as the result. We can write down this mathematically by stating f⁻¹(f(x)) = x. There are a few caveats in regards to what is a true inverse function and weather a function can have an inverse but we won't go into the details here since that would be distracting. Instead we will concentrate on derivatives of inverse functions. It is important to note that when we draw a graph of a function we also draw the shape of the inverse of that function. We know that the inverse function simply accepts different values of y and turns them back into x. Fig. 15 shows how the graph of the function x^3 is just rotated clockwise by 90° and then flipped across the x axis graph to get its inverse function $x^{1/3}$. We know that once we multiply x by itself three times we must take a cube root in order to get back the original x. The cube root is written as $f(x) = x^{1/3}$ and we can algebraically confirm that it is the inverse of $f(x)=x^3$ because the following is true $x = (x^3)^{1/3} = x^{3 \cdot 1/3} = x^1 = x$.

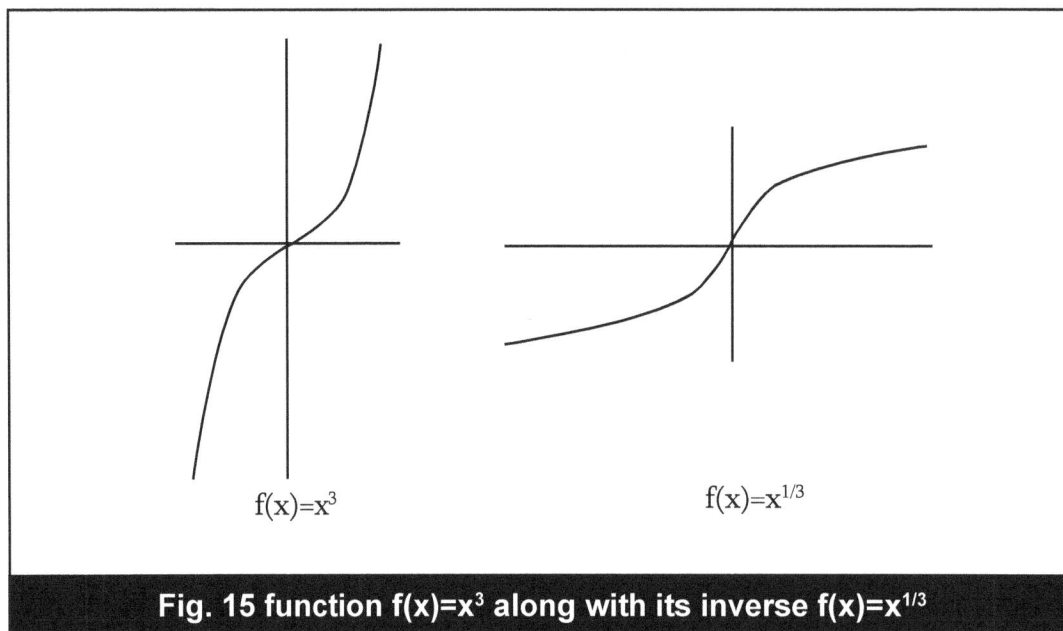

$$f(x)=x^3 \qquad\qquad f(x)=x^{1/3}$$

Fig. 15 function f(x)=x³ along with its inverse f(x)=x^{1/3}

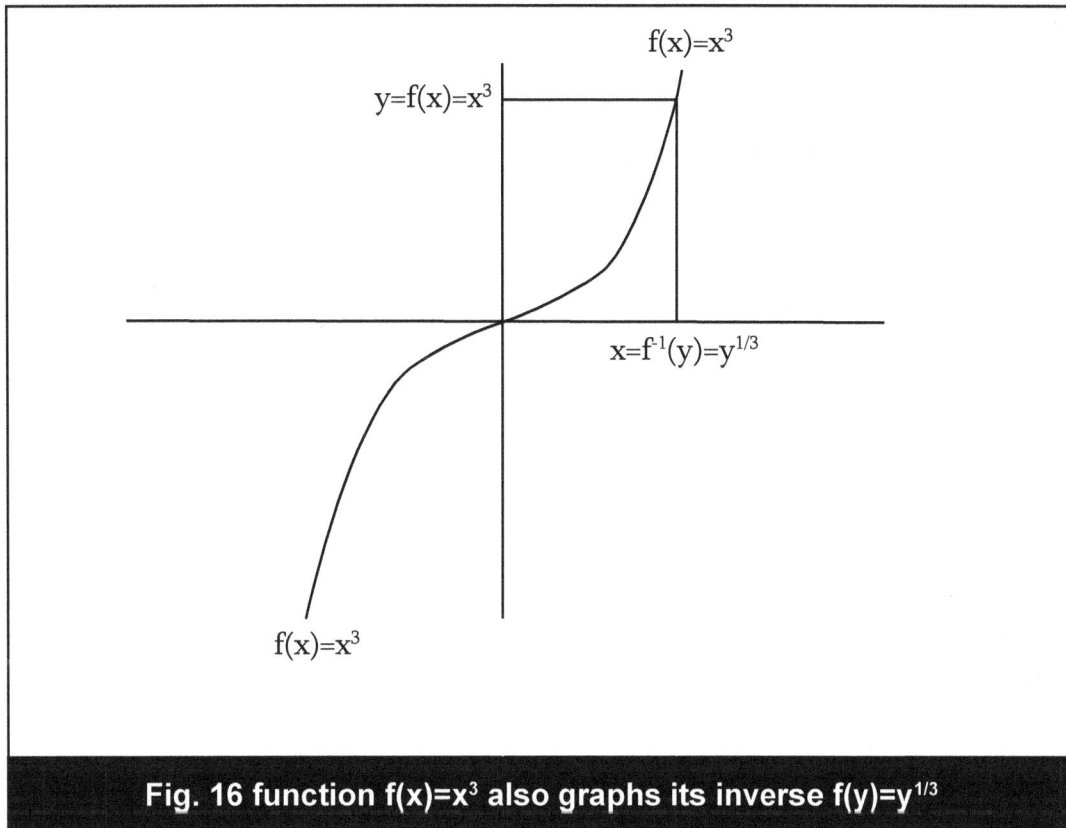

$f(x)=x^3$

$y=f(x)=x^3$

$x=f^{-1}(y)=y^{1/3}$

$f(x)=x^3$

Fig. 16 function f(x)=x³ also graphs its inverse f(y)=y¹ᐟ³

Fig. 16 demonstrates that the function $f(x)=x^{1/3}$ looks like a function $f(x)=x^3$ mirrored across the line y=x.

The function $f^{-1}(y)$ is depicted on every graph of f(x) except that the f(x) axis is the equvalent of its x axis and x axis is its $f^{-1}(y)$ axis. If we know the derivative of the function y=f(x) at some point x then we also know something about the derivative of the inverse function $f^{-1}(y)$ at the point y=f(x). This means that when the derivative of f(x) is depicted in infinitesimally small scale a mirrored image of the derivative of $f^{-1}(y)$ is also visible at the point y=f(x).

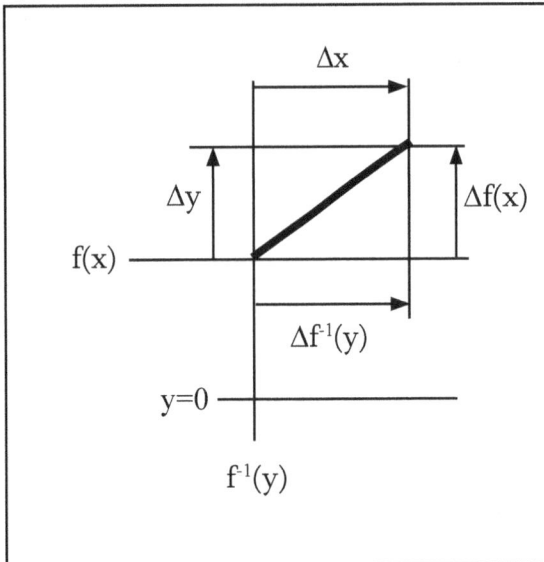

Fig. 17 derivative of f(x)=x³ along with its mirrored derivative of f(x)=x^(1/3)

Fig. 17 clarifies what such a derivative actually looks like. If we zoom into the function we can see that the components that make up the proportions of the derivative of f(x) also make up the proportions of the derivative of the inverse function $f^{-1}(y)$ at the position y=f(x).

Take note that a positive Δy corresponds to the positive $\Delta f(x)$ and that the positive $\Delta f^{-1}(y)$ corresponds to a positive Δx. This means that

$$\Delta f^{-1}(y)/\Delta y = \Delta x/\Delta f(x).$$

If we make the Δx infinitesimally small then the relationship of derivative becomes

$$f^{-1}{}'(y) = 1 / f'(x).$$

Since we know that y = f(x) we can say that

$$f^{-1}{}'(\,f(x)\,) = 1 / f'(x)$$

Which can be rewritten as

$$f'(x) = 1 / f^{-1}{}'(\,f(x)\,)$$

If we substitute the value of the variable x for the value of a new function of x written as m(x) then it can be said that

$$f^{-1}{}'(\,f(\,m(\,x\,)\,)\,) = 1 / f'(\,m(x)\,)$$

If the function m(x) is equal to $f^{-1}(x)$ then it follows that

$$f^{-1}{}'(\,f(\,f^{-1}(\,x\,)\,)\,) = 1 / f'(\,f^{-1}(x)\,)$$

Which is also simplified as

$$f^{-1}{}'(x) = 1 / f'(f^{-1}(x)).$$

To summarize we can just say that

If: f(x) has an inverse $f^{-1}(x)$

Then: $f^{-1}{}'(x) = 1 / f'(f^{-1}(x))$

And also: $f'(x) = 1 / f^{-1}{}'(f(x))$

EXPONENTS

So far we examined addition, multiplication, division and functions of functions. The next logical thing is to understand a derivative of a function which represents a variable raised to some exponent. In order to find the general rule by which we can find a derivative of $f(x) = x^n$ where n is some constant we will examine the cases of n = 1, n = 2 and n = 3. After this examination we will go on to generalize the principles by which derivatives of such functions can be found.

If we start our examination at n = 1 then we have to find the derivative of a function $f(x) = x$. We know that this function represents a graph that intersects the origin and slopes up one unit for each unit of horizontal traversal. For this reason we know that the derivative of this function is f '(x)=1. For our purposes the graph itself would actually be misleading. The graph represents the value of f(x) at various values of x however we are interested in the value of f(x) at a particular instant along with its incremental change. Fig 18 depicts the value of f(x) as a line at a particular value of x.

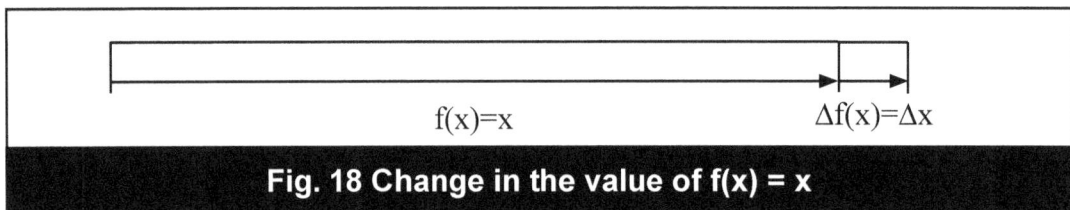

$f(x)=x$ $\Delta f(x)=\Delta x$

Fig. 18 Change in the value of f(x) = x

A miniscule change in f(x) can be represented by an addition to this line. This miniscule addition would have to be divided by the change in x in order for us to know the approximate rate of change. If we wanted to know the derivative at the instant that x is of some value then this proportion $\Delta f(x)/\Delta x$ would have to be examined at that value but the Δx would have to be made infinitesimally small. This proportion would be called the derivative of f(x) at that chosen value of x.

Moving on to the case of $f(x) = x^2$ we can immediately rewrite this as
$f(x) = x * x$. The moment that we do so we can invoke our experience with finding

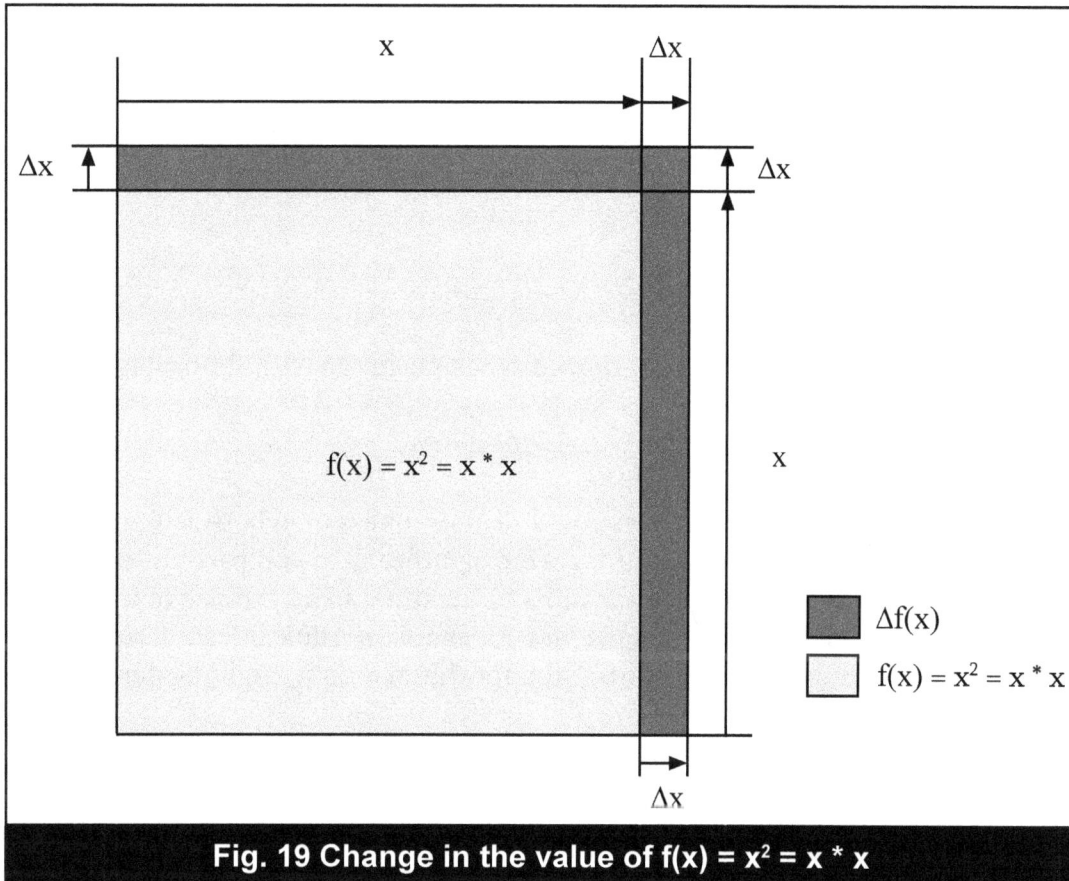

Fig. 19 Change in the value of f(x) = x² = x * x

derivatives of two multiplied functions and represent the instantaneous value of this function as a square in Fig. 19.

If we were to track the growth of this function from x equal to zero to infinity it would form a pyramid with an ever larger base made of squares representing increasing values of x, however this doesn't interest us. We are only interested in the representation of the functions value at one instant of x and how the change in x effects the change in the function value.

Notice that the $\Delta f(x) = \Delta x * x + \Delta x * x + \Delta x * \Delta x = 2 * \Delta x * x + \Delta x * \Delta x$. If we divide both sides by Δx we get

$$\Delta f(x) / \Delta x = 2 * (\Delta x / \Delta x) * x + \Delta x * (\Delta x / \Delta x)$$

which simplifies to

$$\Delta f(x) / \Delta x = 2 * x + \Delta x$$

If we look at the value of this relationship as we make the Δx infinitesimally small we get.

$$f'(x) = 2 * x$$

Notice that the two rectangles composed of only one side whose length was Δx stayed relevant since their proportions were significant in comparison to the size of Δx even as the Δx became really small. On the other hand the size of the square composed of two sides whose length was Δx simply shrunk in size faster in the upper right corner (with Δx shrinking) and for that reason it might be neglected.

Notice that for the case of $f(x) = x$ the value of the function was represented by a one dimensional line and its progression through increasing values of x was represented by a two dimensional graph in two dimensions. In the case where the function is represented by $f(x) = x^2$ we represented the value of the function by a two dimensional square while its progression through increasing values of x was representable by a pyramid in three dimensions. This implies that the representation of the $f(x) = x^3$ will likely be three dimensional in nature and its progression through increasing values of x could be represented in four dimensions.

In order to represent the value of x cubed (hint hint) a cube will be depicted in Fig 20.

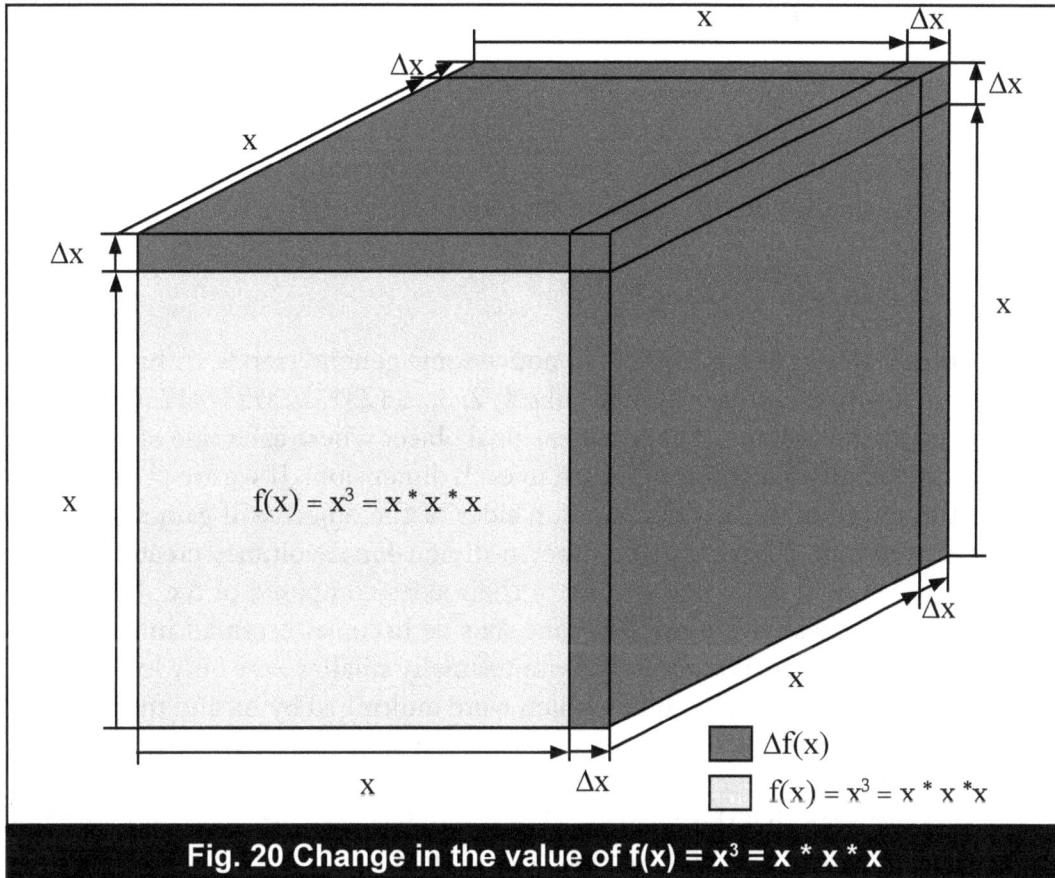

Fig. 20 Change in the value of f(x) = x³ = x * x * x

We can see that the lightly colored inside of the cube is the original value of f(x) and the increase in the size of the cube by Δf(x) is shown in the darker portions of the cube. Each of the three dimensions of the cube are extended by a Δx in order to create the shape that represents the Δf(x). In mathematical terms this visual representation can be stated with an equation.

$$\Delta f(x) = 3 * x * x * \Delta x + 3 * x * \Delta x * \Delta x + \Delta x * \Delta x * \Delta x$$

If we want to find out how Δf(x) changes in relation to Δx we have to divide both sides by Δx.

When we do so we get:

$$\Delta f(x)/\Delta x = 3 * x * x + 3 * x * \Delta x + \Delta x * \Delta x$$

Furthermore if we care to find out what is this rate of change when the Δx is infinitesimally small then the equation simplifies to:

$$f'(x) = 3 x^2$$

At this point it would be appropriate to notice some general trends. A function $f(x) = x^n$ where n is some integer number like 1, 2, 3,..... 1231, ...31241412, etc. can be represented by the volume of an n dimensional object where each side of the object will measure exactly x amount of length in each dimension. If we are to grow such an object by Δx in all n directions then n sides of the object will gain another Δx in their dimension. There will be other n-dimensional volumes created by this change but they will have at least one of their sides composed of Δx. Since all n dimensional volumes with more then one Δx side become too small in relation to Δx when the value of Δx is made to be infinitesimally small we are only left with the n number of n dimensional volumes which were multiplied by Δx and then divided by Δx again when seeking to find the rate at which the function $f(x)$ changes. This means that for intiger values of n = 1, 2, 3, ... 1231,.... 31241412, etc. in the function $f'(x) = x^n$ we can say that the derivative is equal to $f'(x) = n * x^{n-1}$. If you found this explanation to be a bit difficult to comprehend then the following few paragraphs are tailored just for you.

I hope that we established that a function $f(x) = x^n$ where n is some discrete number like 1, 2, 3,..... 1231, ...31241412, etc. can be represented by the volume of an n dimensional object which extends from the origin to the x position along each dimension and all of its sides are of equal x size. The situation gets a little complicated when we try to increase the volume of this object by making each dimension larger by a slight amount Δx. When this takes place we have to consider what are all of the small volumes created that represent the $\Delta f(x)$.

In case of n = 1 we have a one dimensional line the length of x representing the f(x). When we add another Δx to this amount there won't be any other one dimensional "volumes" other then the old x "volume" this Δx "volume" which we just added. It doesn't make sense to talk about x * Δx "volume" since this would create an object with more dimensions then f(x) has.

In case of n = 2 simplicity gives way to complexity. We know that we need to have various combinations of components from side "1" and side "2" which are composed of x and Δx and when combined they will create the whole that we will call $\Delta f(x)$. The x and Δx which belong to side "1" will be written like this x_1 and Δx_1 while the side "2" components will be written like this x_2 and Δx_2. If we take a look at the combination x_1 and x_2 it is immediately recognized that this isn't a part of $\Delta f(x)$ since this piece is the actual f(x) itself. We may also conclude that any n dimensional combination of sides that composes a volume which is entirely composed of x values is the actual function and not a part of the change in the function. Next we examine the two sides x_2 and Δx_2 and we recognize that they actually form a line and not a two dimensional "volume" which is a part of $\Delta f(x)$. This immediately disqualifies the combination x_1 and Δx_1 as well and leads us to a rule that says that any volume composed of the same side and its change in side doesn't have the sufficient number of dimensions to matter in the n dimensional $\Delta f(x)$. After this failed attempt we try x_2 and Δx_1 and we are finally met with success. This combination of sides as well as x_1 and Δx_2 happen to create "volumes" which are a part of $\Delta f(x)$. Notice that the combination of sides x_1 and Δx_2 can be written as Δx_2 and x_1 but it will describe the same volume. The last possibility to create a "volume" is the combination Δx_1 and Δx_2 and indeed we find that this "volume" does show up on our graph.

If we cared to do this in a more formal way we could say something like this. In the case of n=2 we know that there are 3 "volume" types that can exist. One "volume" type where both of the sides of the "volume" are composed of x components. The second "volume" type will have one side composed of an x component and the other will be composed of Δx component. The final "volume" type will be composed of sides which are only Δx components.

Looking at the first type of "volume" we know that we have two x sides to choose from in order to create the two sided "volume". In the two dimensional case it is obvious that this leaves us with only one possibility. The formal process by which we come to this conclusion is as follows. There are two possible sides to choose from to be our first side of the volume. Once we choose our first side we are left with only one other side to choose from in order to compose our two dimensional object. This means that we have 2 * 1 possible "volumes"? But something must be wrong here, how can there be two possible volumes when we know that there is only one. The reason behind this is that our procedure doesn't stop us from choosing x_1 and x_2 but then choosing x_2 and x_1 the next time. In order to eliminate duplicate ways of enumerating the sides of the same "volume" we must ask ourselves how many duplicate ways are there to state all of the sides of a two dimensional object and then divide our result by that number. In this example we know that there are two ways to write down the sides of a two dimensional object but there is a more formal and general way of saying this. We know that we have two sides to work with. If the order in which the sides are written down does matter then we can simply say that for the first side we can choose from two choices (x_1 or x_2) and for the second side we have only one choice (whatever is left x_1 or x_2) and for that reason there are 2 * 1 = 2 different ways to write down the sides of a two sided "volume". Dividing possible "volumes" that we have enumerated by the number of different ways that a two dimensional "volume" can be enumerated we get 2 / 2 = 1. We just discovered that there is only one two dimensional object composed of x sides. This may still be somewhat confusing but the next "volume" type will get us closer to understanding the situation.

The second "volume" type is composed x side and one Δx side. It should be noted that the volume must have the same number of dimensions as the object representing f(x). Furthermore there can't be duplicate uses of the same dimension in the description of the volume or the use of x and Δx of the same dimension in the description of the volume. These properties force us to conclude that if a "volume" isn't composed of the x side of a particular dimension then Δx side of that dimension must be used. Going further with this line of reasoning leads us to conclude that each dimension must be listed and once the x components of the "volume" are picked then the Δx components of other dimensions are also determined. For this reason a "volume" is unlike any other one if its number and selection of x sides is unique. In this case one side is composed of an x side while the other one is composed of an Δx side. There is a choice of two different x sides to choose from for the first and only x side and so we can simply say that there are 2 possibilities. Since we only listed a single side there are no duplicate ways of writing it. For this reason when we look at the number of "volumes" that we can enumerate with one x side and divide it by the number of ways that we can state the same "volume" in our enumeration we get 2/1=2. Simply put there are two "volumes" with one side of x and the other side of Δx.

The third "volume" type is composed of only Δx sides which is a very similar case to the one where the volume was composed of only x sides and for similar reasons we could only conclude that there is only one such volume.

The entire two dimensional case can be summarized down to

$$f(x) + \Delta f(x) = x * x + 2 * \Delta x * x + \Delta x * \Delta x$$

Now it is time for fun in 3D.

In the case where n – 3 we know that there are four types of volumes. By now it should be apparent that there are always n + 1 types of volumes where one type of volume represents f(x) an the other n types represent $\Delta f(x)$.

The first type has all x length sides of different dimensions. We have 3 dimensions to choose from (x_1,x_2,x_3) for the first listed dimension then we have 2 unique dimensions to choose from after that (for example x_2 or x_3 if x_1 is already chosen) until we are finally left with only one (x_3 if x_1 and x_2 are already chosen). This leaves us with 3 * 2 * 1 listed volumes but the volumes are the same regardless of the listed order of their sides. One volume has three sides. There are 3 possible sides that can be listed first then 2 unique sides that can be listed next and finally only one side is left to list. This leaves us with 3 * 2 * 1 ways to list our sides. We divide 3 * 2 * 1 listed volumes by 3 * 2 * 1 ways that a single volume can be duplicated in the list and we are left with 1 volume composed of only x length sides of different dimensions. By now this volume should be recognizable as f(x).

The next type of volume is the one with two x length sides of different dimensions and one Δx length side of a unique dimension too. In order to figure out how many such volumes there are we must consider only how many surfaces with two unique x length sides of different dimensions exist. Once the x surfaces are chosen then Δx sides automatically follow. To do this we must recognize that there are 3 different dimensions of x length sides to chose from first, once chosen there are only 2 other choices of x length sides left for the surface. This means that there are 3 * 2 different x length surfaces that are enumerated. We mustn't forget that we have enumerated surfaces more then once. There are two listed x length dimensions in each surface and this means that for the choice of the first dimension to be listed there are two options to choose from while there is only one other option left to be listed second. That means that there are 2 * 1 ways to denote a single surface. Combining the two effects leads us to conclude that (3 * 2) enumerated surfaces / (2 * 1) ways to duplicate surface enumeration we have 3 truly unique enumerated surfaces.

The type of volume that follows only has one x length side of a unique dimension and two different Δx length sides of unique dimensions. Following the same line of thinking as before we can choose one of the 3 unique x length dimensions for the first dimension. Since there is only one dimension listed there is no way to double its enumeration and we are left to conclude that there is 3 unique volumes with a single unique x length dimension.

Finally we are left with the last volume type which involves only the Δx length sides of unique dimensions. We can explain the fact that there is only one in existence in the same fashion as we did for volumes of x length sides of unique dimensions. Alternatively we can say that since no x length sides present all of the sides are chosen by default and therefore no variations of this type can exist.

This all boils down to

$$f(x) + \Delta f(x) = x * x * x + 3 * x * x * \Delta x + 3 * x * \Delta x * \Delta x + \Delta x * \Delta x * \Delta x$$

In four dimensions where n = 4 this property looks like this.

$$f(x) + \Delta f(x) = 4 * 3 * 2 * 1 / (4 * 3 * 2 * 1) * x * x * x * x$$
$$+ 4 * 3 * 2 * 1 / (3 * 2 * 1) * x * x * x * \Delta x$$
$$+ 4 * 3 / (2 * 1) * x * x * \Delta x * \Delta x$$
$$+ 4 / 1 * x * \Delta x * \Delta x * \Delta x$$
$$+ 0! / 0! * \Delta x * \Delta x * \Delta x * \Delta x$$

which comes out to:
$$f(x) + \Delta f(x) = x^4 + 4 * x^3 * \Delta x + 6 * x^2 * \Delta x^2 + 4 * x * \Delta x^3 + \Delta x^4$$

$$\Delta f(x) = 4 * x^3 * \Delta x + 6 * x^2 * \Delta x^2 + 4 * x * \Delta x^3 + \Delta x^4$$

in five dimensions where n = 5 the formula is

$$f(x) + \Delta f(x) = x^5 + 5 * x^4 * \Delta x + 10 * x^3 * \Delta x^2 + 10 * x^2 * \Delta x^3 + 5 * x * \Delta x^4 + \Delta x^5$$

$$\Delta f(x) = 5 * x^4 * \Delta x + 10 * x^3 * \Delta x^2 + 10 * x^2 * \Delta x^3 + 5 * x * \Delta x^4 + \Delta x^5$$

You should be noticing by now that the first term in $\Delta f(x)$ is always a multiple of n. We formally know there will always be (n * (n-1) * (n-2) * (n-3) .... 4 * 3 * 2 * 1) enumerated "volumes" and ((n-1) * (n-2) * (n-3) ..... 3 * 2 * 1) ways to give duplicate names to those "volumes" which results in n unique "volumes". This is another way of saying that there are only n "surfaces" with unique x length dimensions on the original f(x) "volume". These will be the only "surfaces" composed entirely of x length dimensions which will expand by Δx to compose a part of the greater $\Delta f(x)$ volume.

In any case only the first term which is multiplied by n is left without its Δx when we divide both sides of the equation in order to find the rate of change $\Delta f(x)/\Delta x$. This becomes important because in order to find the derivative we must make the Δx infinitesimally small and that makes all of the other terms too small too quickly to stay relevant. This means that the derivative is defined by the first term of the sum describing $\Delta f(x)$. If the number n is chosen to be an intiger and positive one like 1, 2, ..5, ..8, 291, ...2342.. etc. this term will always be nx^{n-1}. You had to ask for the details, so there they are.

What about the negative exponents? It is pretty hard to imagine objects that have negative dimensions. In order to answer this question properly we must remember that $x^{-n}= 1/x^n$ with this simple property we can find the derivative of x raised to the negative exponent by simply recalling our division example and remembering that where $f(x) = g(x) / r(x)$ it is true that

$$f'(x) = \frac{g'(x)*r(x) - r'(x)*g(x)}{r(x) * r(x)}$$

In this case it would mean that

$$f'(x) = \frac{(0 - n*x^{n-1})}{(x^n * x^n)} = -n / (x * x^n)$$

This simplifies further to

$$f'(x) = -n /x^{n+1} = -n * x^{-n-1}$$

Which leads us to conclude that what was true for positive integers is also true for the negative integers. If number n is an integer then we can say that in functions where $f(x)=x^n$ the derivative is equal to $f'(x)=n*x^{n-1}$.

It is also possible for n to be a fraction of some sort such as r/w. In this case we will write $f(x) = x^{r/w}$. It would be pretty difficult to imagine a fractionally dimensional object, but we won't have to. In this case the function f(x) can be represented as $f(x) = (x^r)^{1/w}$. This reminds us of the case where f(x) was represented by a function of a function. For this reason we can say that $f(x) = g(b(x))$ where $g(b(x)) = b(x)^{1/w}$ and $b(x) = x^r$. The derivative f'(x) is equal to the product of g'(b(x)) and b'(x). The derivative of b(x) is known to be equal to $b'(x) = rx^{r-1}$. In order to find the derivative of $g(b(x)) = b(x)^{1/w}$ we must first remember that the inverse of g(b(x)) is $g^{-1}(b(x)) = b(x)^w$. The derivative of the function $g^{-1}(b(x)) = b(x)^w$ is $g^{-1'}(b(x)) = wb(x)^{w-1}$. Since the inverse function rule tells us that

$$f'(x) = 1 / f^{-1'}(f(x))$$

and in this case that means

$$g'(b(x)) = 1 / (w * (b(x)^{1/w})^{w-1}) = 1 / (w * b(x)^{(1-1/w)})$$

which simplifies further down to

$$g'(b(x)) = (1 / w) * b(x)^{1/w-1} .$$

Multiplying b'(x) by g'(b(x)) where $b(x) = x^r$ we get

$$f'(x) = (r / w) * x^{r-1} * (x^r)^{1/w-1}$$

$$f'(x) = (r / w) * x^{r-1-r+r/w}$$

$$f'(x) = (r / w) * x^{r/w-1}$$

This should prove to us that even when the exponent n is a fraction it is true that

If: $f(x) = x^n$

Then: $f'(x) = n * x^{n-1}$

SIN

When considering the function f(x)=sin(x) the graphical meaning of what that function represents must be considered in order to understand how it changes as the value of x changes.

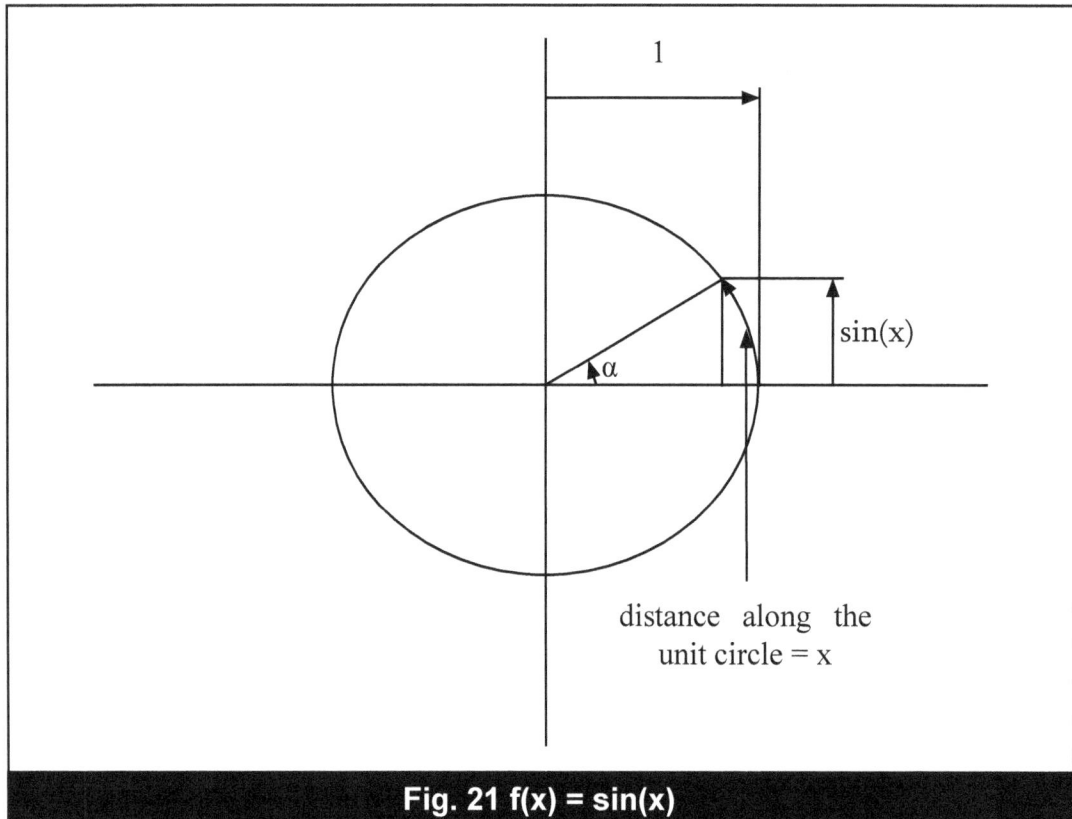

Fig. 21 f(x) = sin(x)

From the Fig. 21 it can be observed that as x increases a path of x length is traversed around the unit circle in the counterclockwise direction starting from its right most position. The length x is equal to the angle α in radians. In order to understand how the travel along this path effects f(x) = sin(x) Fig. 22 will feature a diagram with an exaggerated change in x along with its vertical and horizontal components.

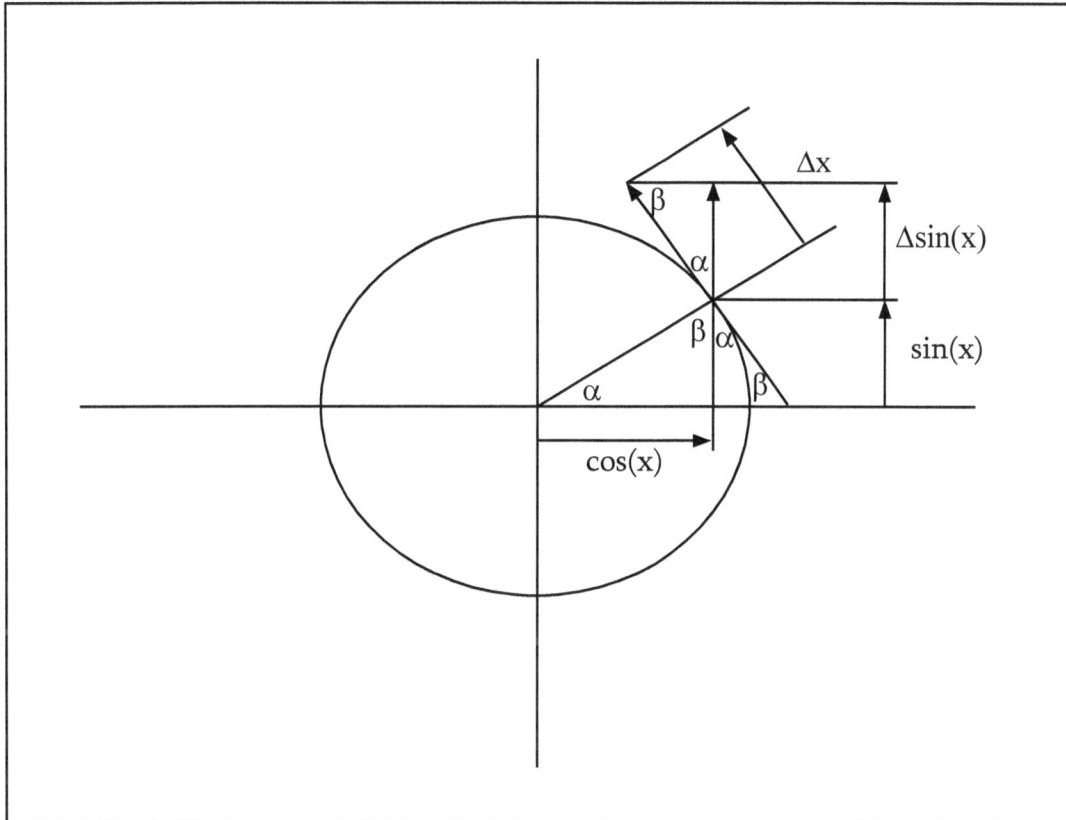

Fig. 22 Δf(x), Δx

In cases where Δx is really small it can be represented by a line tangential to the unit circle at the point at which x currently rests. For the sake of illustrating the geometry the tangential line was exaggerated in Fig. 22. Basic rules of geometry tell us that the triangle describing the relationship between the change in x and the change in the function sin(x) is actually similar to the triangle describing the function sin(x). The difference lies in the fact that the Δsin(x) is actually the cos(x) component of the similar triangle. This means that the change in the function sin(x) will change in proportion to the value of cos(x). But how big does the change get? Is it 5 * cos(x) or maybe 100 * cos(x). The answer to that question lies in the fact that we are actually interested in the relationship Δsin(x)/Δx which becomes the value of the derivative as Δx becomes infinitesimally small. The value of the division can be represented by a similar triangle where the side of Δx can be represented by a side of the value equal to the unit of one. Fig 23. illustrates this in a clearer fashion that allows

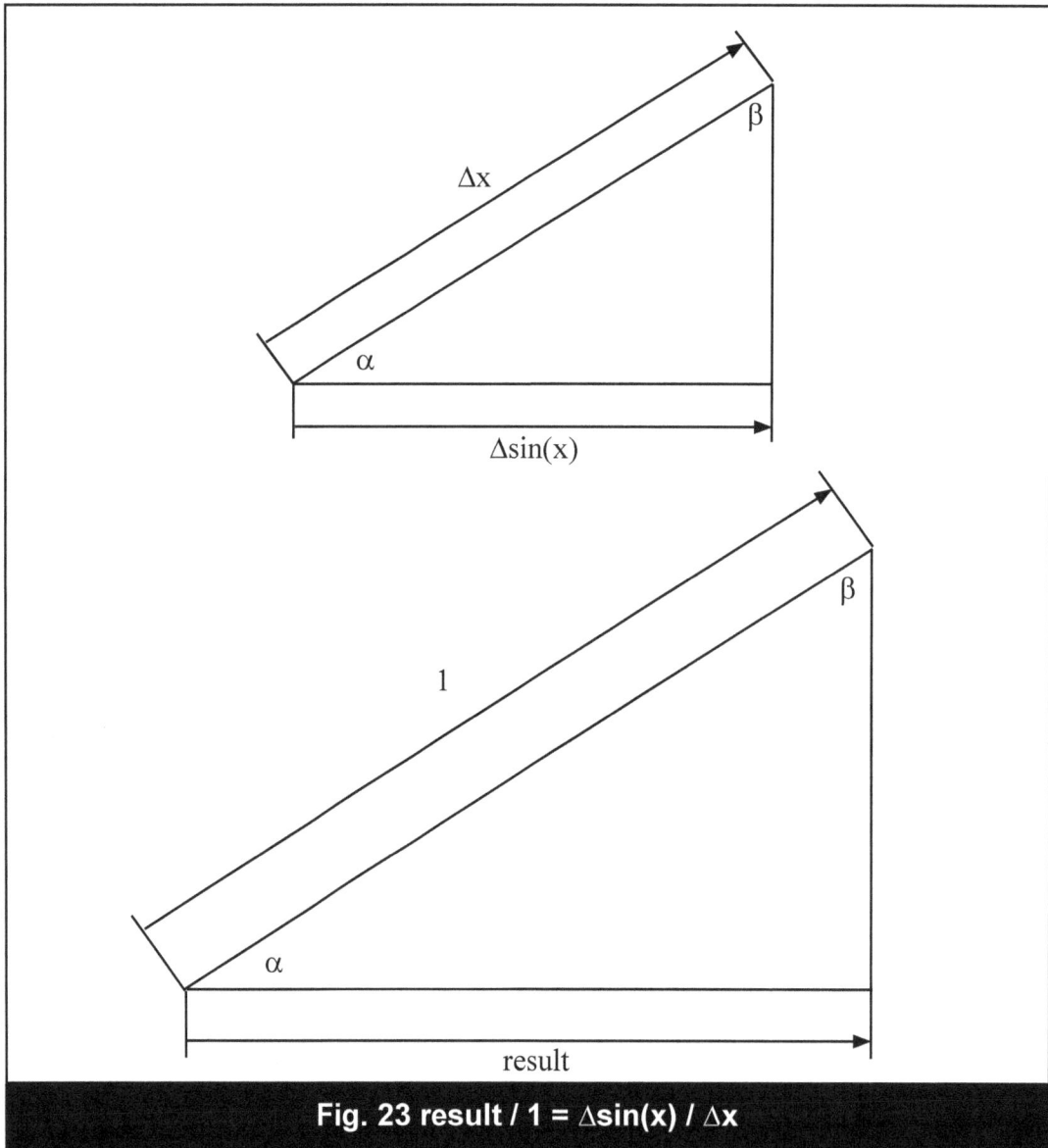

Fig. 23 result / 1 = Δsin(x) / Δx

us to use similar triangles visible in the Fig. 22. You can see two similar triangles with the familiar and friendly angles α and β (where α is in fact x)

Using these similar triangles we can see the proportion $\Delta sin(x)/\Delta x$ = result/1. Such a triangle would always have a right angle and a hypotenuse of 1 so it fit exactly onto the unit circle. Therefore the value representing $\Delta sin(x)/\Delta x$ when Δx is infinitesimally small would always be exactly equal to cos(x). For this reason it can be said that

$$\text{If: } f(x) = sin(x)$$

$$\text{Then: } f\,'(x) = cos(x)$$

COS

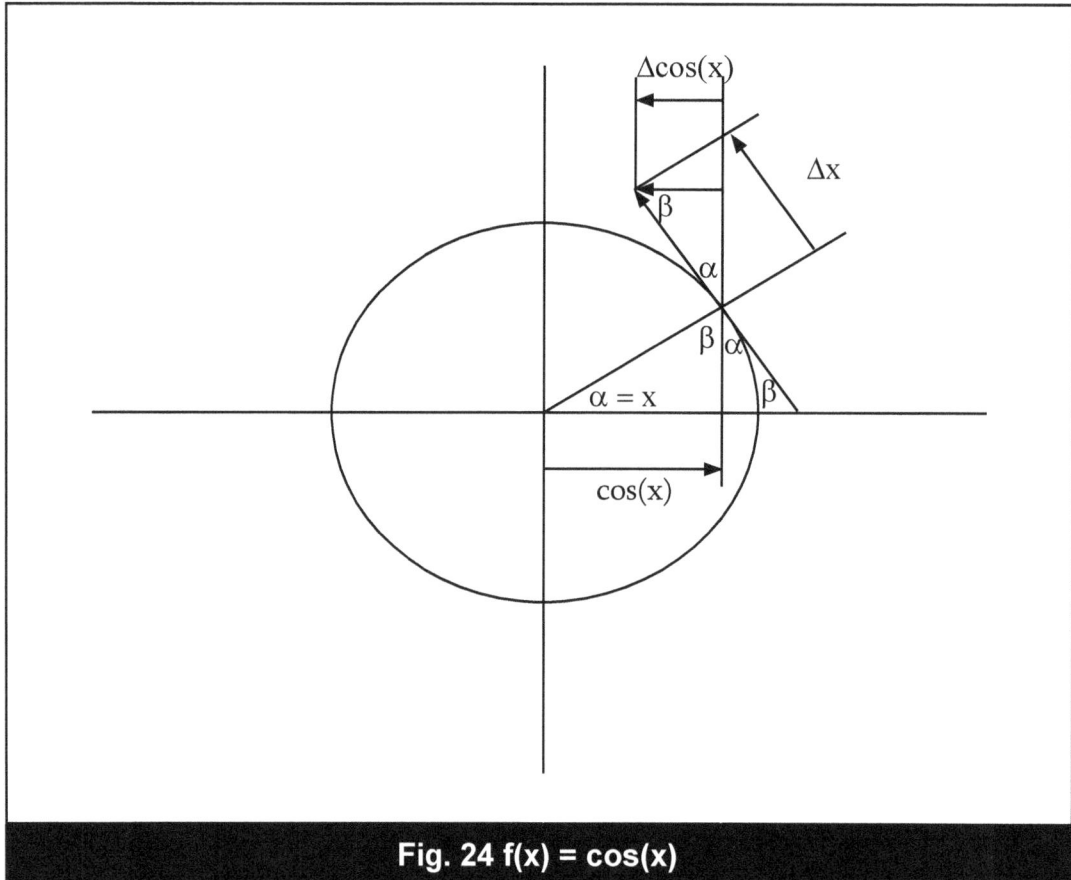

Fig. 24 f(x) = cos(x)

After applying the same methods as in the sin(x) example to analyze the cos(x) a Fig. 24 can be rendered. This graph would imply that the Δcos(x) is proportional in size to sin(x) for the same reasons that Δsin(x) is proportional to cos(x). Something interesting remains to be pointed out however. If you didn't spot it by now you should note that the change is in the negative direction when sin(x) is positive and with further analysis it can be shown that its is positive when the sin(x) is negative.

If all of the parts of the puzzle are pieced together it can be concluded that:

If: $f(x) = \cos(x)$

Then: $f'(x) = -\sin(x)$

TAN

Since we know that $\tan(x) = \sin(x)/\cos(x)$ we can simply use the previously derived rule for division if $f(x) = g(x)/r(x)$ then $f'(x)=(g'(x)r(x)-g(x)r'(x))/r(x)^2$. In this case this results in the equation $(\cos(x)*\cos(x)+\sin(x)*\sin(x)) / \cos(x)^2$.

At this point it would be useful to remember that $\cos(x)$ and $\sin(x)$ are sides of a right angled triangle with a hypotenuse equal to 1. Pythagoras would be pretty upset with us if we didn't realize that $\cos(x)*\cos(x)+\sin(x)*\sin(x) = 1$. With this knowledge in hand we can simply say that

If: $f(x)=\tan(x)$

Then: $f'(x) = 1 / \cos(x)^2 = \sec(x)^2$

CSC SEC COT

These are just instances of functions sin(x), cos(x) and tan(x) raised to the power of -1. Simple application of rules that we already figured out can be used to work out these derivatives. Because this should be embarrassingly simple I refuse to waste your time with the details.

CONCLUSION

Although you were likely never taught why the calculus properties are true, by now you should have a rough idea that all of them can be worked out with a little bit of effort and enthusiasm. The reasons why you were never shown how simple and beautiful calculus really is may vary. They may be as simple and naive as time constraints or perhaps the schooling system is quite disinterested in having you develop a rational and curious mind of your own. But you need not worry about this since I provided you with a few more blank pages to work out these and all the other problems you may face in the future.

www.ingramcontent.com/pod-product-compliance
Lightning Source LLC
Chambersburg PA
CBHW061618210326
41520CB00041B/7492